Albert Spear Hitchcock

A key to the spring flora of Manhattan

Albert Spear Hitchcock

A key to the spring flora of Manhattan

ISBN/EAN: 9783337268534

Printed in Europe, USA, Canada, Australia, Japan

Cover: Foto ©berggeist007 / pixelio.de

More available books at **www.hansebooks.com**

A KEY

TO THE

SPRING FLORA

OF

MANHATTAN.

BY

A. S. HITCHCOCK.

Professor of Botany in the Kansas State Agricultural College,

MANHATTAN, KANSAS.

MANHATTAN, KANSAS

MERCURY PUBLISHING HOUSE.

1894.

PREFACE.

The list includes those Angiosperms in the vicinity of Manhattan which may be found in flower before the first of June. The only Gymnosperm occurring here is JUNIPERUS VIRGINIANA, L.

Through the kindness of Dr. N. L. Britton. I am enabled to present the nomenclature and arrangement as they will probably appear in the new Check List now being prepared by a committee of the A. A. A. S.

Experience has shown that Gray's Manual, though a most excellent work for advanced students, is too extended for the use of the majority of the beginners who can devote but ten weeks to the study of elementary botany and with whom the preparation of a herbarium is subsidiary to a training of the powers of observation.

In giving the characters of orders and genera exceptions are neglected. RANUNCULACEÆ are described as being herbs, that is, they usually are herbs. As the title implies, it is intended to differentiate the plants included in the list only, hence the characters are usually abbreviated and will not serve to definitely distinguish the groups from all others.

ARTIFICIAL KEY TO ORDERS.

CLASS I. Monocotyledoneæ Fibrovascular bundles of stem distributed irregularly through the pith. Venation usually nerved. Parts of the flower often in three's.

Flowers without a perianth, in the axils of small bracts.

A scale above each flower, leaves 2-ranked, sheaths split *Gramineae* 1

No scale above the flower, leaves 3-ranked, sheaths entire, *Cyperaceae* 2

Flowers naked, monœcious, gathered in a fleshy spike surrounded by a spathe.... .... *Araceae* 3

Flowers provided with a proper perianth, divisions 6.

Perianth scale-like.... *Juncaceae* 5

Perianth not scale-like.

Three outer divisions green, 3 inner petal-like.. *Commelinaceae* 4

Divisions similar.

Ovary superior.

Leaves provided with tendrils.... *Smilaceae* 7

No tendrils.... *Liliaceae* 6

Ovary inferior.... *Iridaceae* 8

CLASS II. Dicotyledoneæ. Fibrovascular bundles of stem arranged in a ring. Venation reticulate. Parts of flower usually in four's or five's

SUBCLASS I. Archichlamydeæ. Corolla, if present, of seperate petals.

I. Calyx and Corolla present.

1. Stamens more than ten (cf. Schrankia, which sometimes has more than ten stamens).

Separate.

Inserted on the receptacle.

Herbs.... *Ranunculaceae* 23

Small tree.... *Anonaceae* 22

Woody vine .... *Menispermaceae* 24

Inserted on calyx.... *Rosaceae* 30

United in a column (monadelphous).... *Malvaceae* 43

2. Stamens not more than 10.

As many as petals and opposite them.

Woody vines provided with tendrils.... *Vitaceae* 42

Shrubs, not climbing.... *Rhamnaceae* 41

When as many as petals alternate with them.

A Ovary superior.

Simple; stamens 10; flowers usually papilionaceous.. *Leguminosae* 31

Compound

Stamens as many as petals and alternate.

Herbs, flowers irregular........................ *Violaceae* 44

Woody plants.

Leaves simple........................ *Celastraceae* 37

Leaves compound.

Ovary 1-celled........................ *Anacardiaceae* 36

Ovary 3-celled........................ *Staphyleaceae* 38

Stamens more numerous than petals, or only 2.

Flowers irregular.

Sepals 2; petals 4; stamens 6 .... *Papaveraceae* 25

Petals 4; stamens usually 7 ..... *Hippocastanaceae* 40

Flowers regular.

Stamens 10; petals 5.

Ovary 5-celled.

Leaves simple................ *Geraniaceae* 32

Leaves 3-foliate.............. *Oxalidaceae* 33

Ovary 1-celled................. *Caryophyllaceae* 21

Stamens 6 (or 2); petals 4.

Ovary 1-celled............... *Capparidaceae* 27

Ovary 2-celled...................... *Cruciferae* 26

B. Ovary inferior.

Shrubs.

Leaves opposite; petals 4.................. *Cornaceae* 47

Leaves alternate; petals 5............. *Saxifragaceae* 28

Herbs.

Petals 4.............................. *Oenotheraceae* 45

Petals 5; flowers in umbels.............. *Umbelliferae* 46

II. Corolla and sometimes calyx absent.

Flowers unisexual; one or both sorts in catkins or heads; trees or shrubs.

Staminate flowers in catkins; pistillate single.

Leaves simple.................................. *Fagaceae* 12

Leaves pinnate.............................. *Juglandaceae* 9

Both kinds in catkins or heads.

Ovary 2-ovuled; anthers 1-celled................. *Betulaceae* 11

Ovary 1-ovuled; flowers in heads.............. *Platanaceae* 29

Ovary 1-2-ovuled; flowers in short catkins; anthers 2 celled, *Moraceae* 14

Ovary, many ovuled............................ *Salicaceae* 10

Flowers not in catkins.

Trees or shrubs.

Ovary not lobed.

Leaves simple................................ *Ulmaceae* 13

Leaves pinnate................................ *Oleaceae* 49

Ovary 2-lobed................................ *Aceraceae* 39

Ovary 3-5-lobed; leaves pinnate.................. *Rutaceae* 34

Herbs.

Ovaries several............................ *Ranunculaceae* 23

Ovary 1.
Inferior.................................*Santalaceae* 16
Superior but enclosed in calyx tube, calyx colored, *Nyctaginaceae* 20
Superior, not enclosed in calyx tube.
Ovary 3-celled.........................*Euphorbiaceae* 35
Ovary 1-celled.
Stipules sheathing ....*Polygonaceae* 17
Stipules none.
Flowers closely imbricated with scarious bracts, *Amarantaceae* 19
No scarious bracts.
Flowers perfect; stigmas 2..*Chenopodiaceae* 18
Flowers monœcious or polygamous; stigma 1, *Urticaceae* 15

SUBCLASS II. Sympetalæ. Calyx and corolla both present: petals more or less united.

Ovary inferior.
Flowers in heads surrounded by an involucre
Corollas all strap-shaped ....*Cichoriaceae* 64
Corollas all or partly tubular.... ....*Compositae* 65
Flowers not in heads.
Leaves alternate.............. ....*Campanulaceae* 63
Leaves opposite: corolla 5-lobed.... ....*Caprifoliaceae* 62
Leaves whorled: corolla 4-lobed............ ....*Rubiaceae* 61
Ovary superior.
Stamens as many as lobes of corolla and opposite them. *Primulaceae* 48
Stamens if as many alternate with the lobes of corolla.
Corolla regular.
Ovaries 2: herbs with milky juice.
Stamens distinct.... ... .*Apocynaceae* 50
Stamens united with each other and the stigma. *Asclepiadaceae* 51
Ovary 1.
Deeply 4-lobed around the style ....*Asperifoliae* 55
Not lobed.
Leaves opposite, entire: ovary 3-celled. *Polemoniaceae* 53
Leaves opposite, entire or toothed; ovary 2-celled, *Scrophulariaceae* 59
Leaves all radical; flowers in a close spike, *Plantaginaceae* 60
Leaves alternate, or if opposite, deeply parted, not all radical.
Stamens 10 or more.. ....*Leguminosae* 31
Stamens 5.
Ovules not more than 4.
Ovary 2-celled: leaves not lobed. *Convolvulaceae* 52

Ovary 1-celled; leaves deeply parted, *Hydrophyllaceae* 54
Ovules numerous; ovary 2-celled. *Solanaceae* 58
Corolla irregular.
Ovary deeply 4-lobed around the style... .... *Labiatae* 57
Ovary not lobed.
2-celled; ovules numerous............ *Scrophulariaceae* 59
2-4-celled; ovules 4........... .. .. *Verbenaceae* 56

DESCRIPTIVE LIST.

MONOCOTYLEDONEAE.

1 GRAMINEÆ.

Perianth none. Stamens usually 3. Ovary one, superior, 1-celled, 1-ovuled. Styles and stigmas 2. Fruit a caryopsis, the seed grown fast to the pericarp. Flower in the axil of a bract (flowering glume) with a 2-keeled bract (palet) between the flower and the axis. The flowers are arranged in 2 ranked clusters (spikelets), the two glumes at the base of the spikelet being empty. The spikelets may be from one to many flowered. Herbaceous plants with 2-ranked, alternate, nerved leaves and sheaths split on the side opposite the blade

1. Spikelets with usually 4 glumes and one perfect flower, jointed upon the pedicel below the glumes
   Spikelets in loose panicles.......... ....*Panicum*
   Spikelets in spike-like panicles. ... .*Chamaeraphis*
2. Spikelets not jointed below the glumes, 1-many flowered; glumes 3 when 1-flowered.
   a Spikelets 1-flowered, not on a zig-zag rhachis.
   Inflorescence loosely panicled... ....*Agrostis*
   Inflorescence spike-like................................*Alopecurus*
   b. Spikelets diœcious; staminate in two rows, forming a 1-sided spike.. ....*Bulbilis*
   c Spikelets more than 1-flowered, pedicelled.
   *Flowering glume 3-nerved.
   Empty glumes similar, acute.... ....*Kœleria*
   Empty glumes very dissimilar, upper very obtuse ....*Eatonia*
   * *Flowering glume 5-nerved
   Glumes compressed and keeled.
   Awn-pointed; spikelets in 1-sided clusters.... ....*Dactylis*
   Not awn-pointed: spikelets in panicles.....................*Poa*
   Glumes convex on back.
   Awn, if present, from tip of glume...... .... *Festuca*

Glumes awned below the 2-cleft apex.............. .. ...*Bromus*

d. Spikelets sessile on opposite sides of a zig-zag rhachis.

1-flowered........... ....*Hordeum*

3-5-flowered..... ... .*Elymus*

PANICUM, L.

Lower glume usually small; third glume empty or staminate: fourth glume coriaceous, enclosing a perfect flower.

P laxiflorum, Lam. Resembling the next but spikelets scarcely 2 mm. long. Prairie land

P. scoparium, Lam. Stems usually several, 2-4 dm. high Sheaths hairy. Panicle about ½dm. long Spikelets obovate, obtuse, 3 mm. long; lower glume roundish, about one-third the length of the spikelet. Prairie land, common.

CHAMÆRAPHIS, R. Br.*

Spikelets as in *Panicum* but the peduncles bearing a few long bristles below the joint.

C. viridis, (L.) Porter. Annual, upright; leaves straight (not twisted as in *C glauca*). Spike 3-10 cm. long, usually somewhat tapering upward. A common weed in cultivated ground.

ALOPECURUS, L.

Spikelets 1-flowered; lower glumes boat shaped, about as long as the flowering glume which is awned on the back below the middle. Inflorescence spike-like

A. geniculatus, L. Tufted, 2-4 dm. high, leaves short, the soft spike 2-5 cm long, often partly enclosed in the upper sheath. Moist meadows, frequent.

AGROSTIS, L

Spikelets 1-flowered. Empty glumes usually longer than the flowering one, all without terminal awns.

A. hiemalis, (Walt) B. S. P. Slender stems tufted, with a few short leaves at base, 3-4 dm high Panicle large, the branches capillary and in fruit very long and spreading. Lower glume acute, about 1½ mm. long. Dry land, common.

BULBILIS, Raf.

Diœcious; staminate spikelets 2-ranked in 1-sided short spikes, 2-3 flowered; spikes 1-3 at the summit of the short stem. Pistillate spikelets 1-flowered in a cluster near the ground, each in the axil of a leaf-like bract.

B. dactyloides, (Nutt.) Raf. A low tufted grass, spreading extensively by stolons. Prairie, frequent.

EATONIA, Raf

Spikelets 2 flowered; empty glumes nearly equal in length but very unequal in shape, lower narrow, upper broad and obovate.

---

*It would seem best to retain *Setaria* as a genus distinct from the Australian *Chamaeraphis*, but as the name was used earlier for a genus of lichens, a new name should be given.

F. obtusata, (Michx.) Gray. An upright smooth grass ½ to ⅔ m. high, with contracted panicle. Prairie, common.

KŒLERIA, Pers.

Spikelets 2-4-flowered; empty glumes somewhat unequal, about the length of the spikelet, acutish.

K. cristata, (L.) Pers. Resembling the previous species but stem pubescent above and panicle rather more contracted. Prairie, common.

DACTYLIS, L.

Spikelets 3-4-flowered, crowded in 1-sided clusters; empty glumes short awned, flowering glume 5 nerved.

D. glomerata, L. An upright, tufted grass ½ m. high or more, with rather broad leaves. Escaped from cultivation.

POA, L.

Spikelets ovate, compressed, 2-several flowered, in open panicles. Empty glumes shorter than flowers. Flowering glume keeled, 5 nerved, with a scarious margin. Awns none.

P. compressa, L. Panicle short and narrow. Stem flattened, wiry, 2-4 dm. high. Leaves short and erect. Perennial by running rootstocks. Sterile ground.

P. pratensis, L. Panicle pyramidal, short. Stem cylindrical, upright, 2-6 dm. high. Escaped from cultivation, common.

P. sylvestris, Gray. Panicle loose and open, pedicels slender. Stem (2-6 dm.) and leaves soft and weak. Upland woods.

FESTUCA, L.

Spikelets 3-several flowered. Panicle open or contracted. Flowering glume coriaceous, convex on back, 3-5 nerved, acute or awned from tip.

F. octoflora, Walt. Panicle often contracted into a simple raceme. Flowering glume short awned. Leaves few and short, convolute. Stem erect, 1-4 dm. high. Sterile soil, common.

F. elatior, var. *pratensis*, (Huds.) Hack. Panicle simple. Glumes awnless. Leaves flat. Stem tall; ½ m. or more. Escaped from cultivation.

BROMUS, L.

Spikelets several flowered, panicled. Empty glumes unequal, nerved. Flowering glume convex on back, or keeled above, nerved, awned below the 2-cleft apex.

B. secalinus, L. Panicle often simple, pedicels rough, slender. Lower empty glume acute, narrow, 3-nerved; upper broader and longer, 7-nerved. Flowering glume about 7-nerved; awn shorter than glume. A weed in fields.

HORDEUM, L.

Spikelets 1-flowered, sessile on a zig zag rhachis, 3 at each joint, the 2 lateral more or less imperfect, the 6 empty glumes bristle form, standing side by side in front of the spikelets. Flowering glume long awned from the apex.

H. jubatum, L. Awn capillary, spreading, 2-5 cm. long. Stems about ½ m. high. Waste places, infrequent.

H. nodosum, L. Awn about as long as glume (½ cm.) erect. Stem low, 1-5 dm. Spike sometimes enclosed in upper sheath, breaking up at maturity into joints. Sterile soil, common.

ELYMUS, L.

Much as in *Hordeum* but spikelets all perfect and usually more than 1-flowered.

E. Canadensis var. *glaucifolius*, (Willd.) Torr. Spike large and thick. Empty glumes strongly nerved, awned. Flowering glumes with long spreading, capillary awns. Large grass with rough leaves. Low land, common.

2. CYPERACEAE.

Proper perianth none, sometimes represented by bristles or scales. Stamens usually 3. Ovary one, superior, 1-celled, 1-ovuled. Style 2-3-cleft Fruit an achene. Flowers in the axils of bracts; no bract between the flower and the axis. Grass-like plants with 3-ranked leaves and closed sheaths.

ELEOCHARIS, R. Br.

Flowers perfect. in a single spike terminating the naked stem. Ovary surrounded by several bristles. Style thickened at the base forming a tubercle on the achene.

E. palustris, (L.) R. & S. Stems cylindrical, striate, 1-5 dm. high. In mud or shallow water, common.

SCIRPUS, L.

Resembling *Eleocharis* but style not thickened. In our species the 1-several spikelets are at the apex of the stem, with an involucral leaf which may appear as a continuation of the stem.

S. Americanus, Pers. Spikelets 1-several, sessile, involucral leaf much longer than the cluster. Stems triangular. Wet places.

S. lacustris, L. Spikelets panicled, the involucral leaf shorter than the cluster. Stems cylindrical. tall. Wet places.

CAREX, L.

Flowers monœcious, the two kinds in the same or different spikes. Achene enclosed in a sac (perigynium). Stems usually triangular.

§ 1. Staminate flowers in one or more terminal spikes; the pistillate spikes below, usually peduncled. Achene triangular.

C. hystricina, Muhl. Pistillate spikes 1-3, drooping on slender stalks, oblong and densely flowered. Perigynium smooth, strongly nerved, somewhat inflated, narrowed into a slender beak as long as the body. Swampy ground.

C. filiformis. var. *lanuginosa*, (Michx.) B. S. P. Pistillate spikes 1-3, erect and nearly sessile. Perigynium short beaked, deeply 2-toothed. densely pubescent. Low prairie.

C. trichocarpa, Muhl. Staminate spikes several. Pistillate spikes usually 2-3, upper erect, lowermost long stalked and spreading. Perigynium smooth, nerved, the short beak extending into 2 long, bristle-like teeth. Coarse plants growing in marshes.

C. stricta, Lam. Spikes several: pistillate slender and compactly flowered, mostly several and erect. Perigynium short, smooth, scarcely nerved, beak short and entire Swampy land.

C. tetanica, Schk. Staminate spike one; pistillate spikes 1-3, short, more or less stalked. Perigynium nerved, smooth, the very short beak bent to one side. Prairie, common. Leaf below lowermost spike several cm. long.

C. Pennsylvanica, Lam. The habit of the preceding but perigynium pubescent, beak longer and toothed, and lower leaf only a little longer than spike. Prairie, common.

§ 2. The two kinds of flowers in the same spike. Spikes short, sessile and collected at the summit of stem. Perigynium plano-convex and achene lenticular.

C. Muhlenbergii, Schk. Staminate flowers borne at top of spike, hence lower scales fruitful Perigynia strongly spreading at maturity. Spikes close together, so as to form an aggregate spike-like head. Prairie, common.

C. straminea, var. *festucacea*, (Willd.) Tuckerm. Staminate flowers at base of spike, hence several of the lower scales empty causing the spike to taper below. Perigynia ascending Spikes usually 4-6, shortly separated from each other. Common on prairie

3. ARACEAE.

Flowers sessile, crowded on a fleshy axis (spadix). Fruit a berry.

ARISÆMA, Mart.

Flowers naked, covering only the lower portion of the spadix, the latter surrounded by a rolled up leaf or spathe Perennial from a corm. Leaves compound. Fruit scarlet, 1-few seeded.

A. Dracontium, (L.) Schott. Usually one leaf with several leaflets. Spadix tapering into a long slender point which is exerted from the rolled up tip of the spathe. Low woods, frequent,

A triphyllum, (L.) Torr. Leaves 2, each with 3 elliptical leaflets. Spadices diœcious, club-shaped and obtuse above, shorter than the hooded spathe Rich woods, rare.

4. COMMELINACEAE.

Calyx of 3 green sepals. Corolla of 3 colored petals which soon wither. Stamens 6. Ovary 2-3-celled, free. Fruit a capsule.

TRADESCANTIA, L.

Petals pink or violet. Filaments covered with long violet hairs. Flowers umbelled in terminal and axillary clusters.

T. Virginiana, L Common in low prairie. Flowers opening on sunny mornings and soon withering.

5. JUNCACEAE.

Perianth of 6 similar, persistent, bract-like divisions Stamens 6. Ovary one, superior, 3-carpelled. Style 1, stigmas 3. Fruit a 3 valved capsule. Grass-like herbs

JUNCUS, L

J. tenuis, Willd. Stem wiry, 2-4 dm. high. Flowers small, green.

½ cm. long, in terminal panicles. Sepals narrow, acute. Common in grass land.

6. LILIACEAE.

Perianth of 6 similar divisions, not bract-like. Stamens 6 Ovary free, 3-celled, several ovuled. Fruit a capsule or berry.

Flowers borne on a leafless stem.
  Solitary and nodding; from a corm ....*Erythronium.*
  Umbelled; from a bulb
    The bruised plant exhaling the odor of onions............*Allium*
    No odor of onions ....*Nothoscordum*
  Racemose or panicled: from a thick woody rootstock ....*Yucca*
Stems leafy.
  Leaves scale-like with thread-like branches in their axils. *Asparagus*
  Leaves broad.
    Flower clusters axillary...............................*Polygonatum*
    Flower clusters terminal
      Leaves grass like: from bulbs.........................*Zygadenus*
      Leaves elliptical; from rootstocks.......................*Vagnera*

ZYGADENUS. Michx.

Z. Nuttallii, (Gray) Wats. Flowers yellowish, in racemes. Stem 4-8 dm. high. Rocky hills, frequent.

ALLIUM, L.

A mutabile, Michx 2-4 dm. high from a densely fibrous coated bulb. Leaves about half the length of scape Low prairie.

NOTHOSCORDUM, Kunth.

N. ornithogaloides. (Walt.) Kunth. Leaves about the length of the scape. Low prairie.

ERYTHRONIUM, L.

Scape about the length of the two smooth, elliptical leaves

E albidum, Nutt. Divisions of perianth white or pinkish, 2-3 cm. long, recurved. Low woods, infrequent

YUCCA, L

Leaves long and rigid, linear, sharp pointed Flowers white, large and showy.

Y. glauca, Nutt. Flowers 4-6 cm. broad, in an elongated raceme Sand hills, not common.

ASPARAGUS, L.

Stems much branched, the filiform branches performing the function of leaves. Fruit a scarlet berry.

A. officinalis, L Escaped from gardens.

VAGNERA, Adans.

Flowers small, white. Stems simple.

V. racemosa, (L.) Morong. Perianth divisions 2 mm. long. Flowers in panicles. Stem pubescent. Rich woods, rare.

V. stellata, (L.) Morong. Perianth divisions about 4 mm. long. Flowers in racemes. Stem smooth. Rich woods, rare

POLYGONATUM, Adans.

Perianth greenish, cylindrical, united, 6-lobed at summit. Stamens inserted on perianth. Fruit a globose berry. Stem from a creeping fleshy rootstock.

P. biflorum, var. *commutatum*, (R. & S.) Morong. Stem ½-1 m. high, naked below. Peduncles axillary, few flowered. Rich woods.

7. SMILACEAE.

SMILAX, L.

Flowers diœcious, small and greenish, in peduncled axillary umbels. Fruit a black few seeded berry. Leaves netted veined, their petiole bearing two tendrils.

S. hispida, Muhl. A woody climbing vine; the stem, at least below, armed with numerous black prickles. Low woods, common.

8. IRIDACEAE.

Divisions of perianth 6, colored. Stamens 3. Ovary inferior, 3-celled. Leaves equitant.

SISYRINCHIUM, L.

Stamens monadelphous. Leaves grass-like. Roots fibrous.

S. Bermudianum, L. Flower stem flat, 1-2 dm. high, bearing a few slender pedicelled white or blue flowers. Prairie, common.

DICOTYLEDONEAE.

9. JUGLANDACEAE.

Flowers monœcious, the staminate in lateral catkins, the pistillate single or in clusters terminating the growth of the season. Trees with pinnate leaves.

JUGLANS, L.

Fruit with an indehiscent spongy covering. Pith diaphragmed.

J. nigra, L. Leaflets 7-11 pairs. Twigs downy. Low woods, frequent.

HICORIA, Raf.

Covering of fruit splitting from apex into four parts. Pith without diaphragms.

H. ovata, (Mill.) Britton Leaflets mostly 5, lower pair much smaller. Bud scales several, Bluffs along river below Manhattan.

H. minima, (Marsh) Britton. Leaflets mostly 7-9. Bud scales 2, yellow. Low woods, common.

10. SALICACEAE.

Flowers diœcious, both kinds in catkins. Trees or shrubs with alternate, simple leaves.

POPULUS, L.

Scales of catkins cut-lobed. Flowers from a cup-shaped disk. Trees with broad leaves.

P. monilifera, Ait Leaves broadly ovate with a slender point, on a flattened, slender petiole, serrate. Low ground, common.

SALIX, L

Scales of catkins entire. No disks. Leaves long and pointed, ours with serrate leaves.

S amygdaloides, Anders. Leaves pale beneath, closely and sharply serrate. Catkins terminating small leafy branches of the season. Stamens 3 or more. Wet places, becoming a large tree, common.

S longifolia, Muhl. Leaves narrow, remotely denticulate. Stamens 2. A shrub or tree, common in low places especially along streams.

S. cordata, Muhl. Leaves glaucous beneath, closely serrate. Flowers appearing before the leaves. Stamens 2. A tall shrub. Rocky ravines, not common.

11. BETULACEAE.

Flowers monœcious, the staminate in catkins. Ovary 2-celled, 2-ovuled. Fruit a 1-seeded nut. Woody plants with alternate, simple leaves.

OSTRYA, Scop.

Staminate flowers in close, cylindrical, bracted catkins, from the previous season's wood. The pistillate in short catkins terminating the growth of the season. Fruit an achene, enclosed in an inflated, flattened sac. Leaves 2-ranked.

O Virginiana, (Mill) B. S. P. Leaves oblong, doubly serrate, pinnately veined. Bark twisted. Bluffs, frequent.

12. FAGACEAE.

Differs from *Betulaceæ* in having the ovary 3-celled and 3 or 6 ovuled.

QUERCUS, L.

Staminate catkins very slender, interrupted, bractless, from the old wood. Pistillate flowers in little clusters along the new growth. Fruit an acorn. Leaves 5-ranked.

Q. macrocarpa, Michx. Bark light colored; twigs corky ridged. Leaves sinuate-pinnatifid and lobed. Acorns nearly covered by the large mossy fringed cup. Common.

Q. prinoides, Willd. Bark light, twigs smooth. Leaves irregularly sinuate-toothed and pinnately veined. Acorns small. Upland woods, common.

Q. nigra, L. Bark dark, twigs brown. Leaves thickish, 3-lobed at apex, or often lobed along the sides; lobes bristle pointed. Acorn small, about half enclosed in the hemispherical, coarsely scaly cup. Dry hills, from Manhattan eastward.

Q Rubra, L. Bark dark and smooth, twigs smooth. Leaves lobed, the lobes bristle pointed. Acorn large with a very flat, shallow cup. Hills east of Manhattan, frequent.

Q tinctoria, Bartr. Resembles *Q. rubra* but acorn with a top-shaped or hemispherical cup covering the lower half of the acorn. Hills east of Manhattan.

13. ULMACEAE.

Flowers more or less polygamous, in umbels or racemes, not in catkins. Trees with alternate, 2-ranked, pinnately veined leaves. Calyx free from the 1-2-celled ovary.

ULMUS, L.

Flowers often perfect. Fruit a 1-2-celled samara, winged all around, which ripens in early spring. Flowers appearing before the leaves, in clusters on the old wood.

U. Americana, L. Flowers slender pedicelled. Fruit hairy margined. Leaves smooth above. Twigs brown, smooth or nearly so. Woods, common.

U. pubescens, Walt. Flowers nearly sessile. Fruit smooth. Leaves rough above. Twigs gray, scabrous. Woods, common.

CELTIS, L

Flowers racemose from the growth of the season. Fruit a drupe.

C. occidentalis, L. Leaves taper-pointed, nearly smooth, veiny. Bark gray and rough-warty. Low woods, frequent.

14. MORACEAE.

Flowers unisexual. Calyx becoming fleshy in fruit, enclosing the achene. Woody plants with milky juice.

MORUS, L.

Flowers in catkin-like spikes; the fertile spike resembling in fruit a blackberry.

M. rubra, L. Leaves cordate, serrate, often deeply lobed. Twigs smooth, light gray. Low woods, frequent.

15. URTICACEAE.

Flowers unisexual. Ovary 1-celled. Fruit an achene. Herbs.

PARIETARIA, L.

Flowers in bracted cymose clusters in the axils of the alternate, entire, 3-ribbed leaves.

P. Pennsylvanica, Muhl. A low, simple stemmed, inconspicuous annual, with thin, roughish leaves. Moist, shady places, common.

16. SANTALACEAE.

Flowers apetalous. Calyx 4-5-cleft, adherent to the 1-celled ovary. Stamens as many, inserted opposite the lobes. Fruit 1 seeded. Style one. Leaves entire.

COMANDRA, Nutt.

Low perennials with alternate, sessile leaves and whitish flowers in terminal clusters.

C. umbellata, (L.) Nutt Stem erect (1-2 dm). Leaves oblong. Prairie, common.

17. POLYOGONACEAE.

Flowers apetalous. Calyx mostly 5-6-cleft, persistent. Stamens several. Ovary 1-celled, 1-ovuled, forming in fruit an achene. Styles 2-3. Herbs with alternate, entire leaves and sheathing stipules.

RUMEX, L.

Sepals 6, the 3 outer reflexed in fruit, the inner enlarging and enclosing the achene, then called *valves*. Stamens 6. Achene 3-angled. Flowers small, green, in close whorls, forming a terminal panicle.

R. Acetosella, L Low and slender, diœcious. Leaves halberd-shaped or the upper linear. Valves small. Meadows, infrequent.

R. altissimus, Wood. Taller (½-1 m). Leaves oblong-lanceolate, not wavy. Valves veiny (4 mm. broad), one with a conspicuous thickening. Low ground, common.

R. crispus, L. Resembles the foregoing but the leaves with a conspicuously wavy margin. Low ground, common.

R. venosus, Pursh. Low (2-3 dm.) Leaves oblong, with conspicuous stipules. Valves very large (2-3 cm.), orbicular, pink. Low ground, infrequent.

POLYGONUM, L.

Calyx mostly 5 parted, all the lobes erect in fruit. Styles 2 or 3 and the achene lenticular or triangular.

P. aviculare, L. Low and slender annual, prostrate or in wet places upright. Leaves small. Flowers axillary. Stamens 8. Achene triangular. Along roadsides, etc., common.

P. Convolvulus, L. Twining annual. Leaves halberd-shaped. Flowers in racemes. Achene triangular. Fields, common.

P. Pennsylvanicum, L. Erect annual. Leaves lanceolate. Peduncles glandular. Flowers pink or white in terminal spikes. Achene lenticular. Moist soil, common.

P. Persicaria, L. Differs from the above in having the leaves often marked with a dark spot, the sheaths with ciliate margin, peduncles smooth and achene sometimes triangular. Wet places, common.

18. CHENOPODIACEAE.

Flowers apetalous, small and green; no bracts. Stamens as many as sepals, inserted opposite them. Ovary 1-celled, 1-ovuled. Fruit an utricle. Herbs with mostly alternate leaves and no stipules.

CHENOPODIUM, L.

Sepals 5. Flowers sessile in small clusters. Plants usually mealy. Ours annual.

C. album, L. Leaves ovate, more or less angulate-toothed. A common weed

C. leptophyllum, Nutt. Leaves linear, entire. Dry places, common.

19. AMARANTACEAE.

Flowers imbricated with dry bracts, about 3 to each flower. Ovary 1-celled, 1-ovuled. Herbs without stipules.

AMARANTUS, L.

Sepals mostly 5. Stamens 5. Fruit an utricle. Leaves alternate, entire. Flowers green, very small.

A blitoides, Wats. Prostrate. Leaves obovate. Roadsides, etc. Common in summer and autumn.

20. NYCTAGINACEAE.

Calyx gamosepalous, colored like a corolla, the base closely surrounding the 1-celled, 1-ovuled ovary. Corolla none. Herbs with opposite entire leaves.

ALLIONIA, L.

Flowers 3-5 in an open involucre. Calyx pink.

A. nyctaginea, Michx. Leaves ovate, petioled. Stem ½-1 m. high. Dry places, frequent.

21. CARYOPHYLLACEAE.

Herbs with opposite, entire leaves. Stamens not more than twice the sepals. Ovary 1-celled with central placenta.

SILENE, L.

Calyx 5-toothed. Stamens 10. Styles 3.

S antirrhina, L. A slender annual with linear leaves and small flowers on long peduncles, each of which is provided with a glutinous ring. Dry ground, common.

S. vulgaris, (Moench), Garke. Leaves ovate-lanceolate Calyx inflated, veiny. Large petals 2-cleft. Escaped occasionally.

22. ANONACEAE.

Sepals 3. Petals 6. Stamens numerous. Pistils several. All hypogynous. Woody plants.

ASIMINA, Adans.

Pistils ripening into 1 or 2 oblong, yellow, pulpy fruits.

A. triloba, (L) Dunal. A small tree with oblanceolate leaves and dark brownish-red flowers. Fruit edible.

23. RANUNCULACEAE.

Parts of flower all free and distinct. Sepals often corolla-like. Petals often absent. Stamens numerous. Pistils few to several, 1-celled. Fruit an achene, follicle or berry. Herbs.

Flowers diœcious, small, panicled. ....*Thalictrum*
Flowers perfect.
  Not spurred. Fruit an achene.
    Petals absent..*Anemone*
    Petals present.
      Achenes in a long spike............................*Myosurus*
      Achenes in a head.....................................*Ranunculus*
  Conspicuously spurred. Fruit a follicle.
      Spur one........... .. ..*Delphinium*
      Spurs 5..*Aquilegia*

THALICTRUM, L.

Petals none. Fruit an achene.

T. purpurascens, L. Stem 1 m. high. Leaves decompound. Achenes ribbed. Low ground, infrequent

ANEMONE, L

Leaves radical; the stem leaves forming a 2-3 leaved involucre on each peduncle.

A. Caroliniana, Walt Stem 1-2 dm. high. Leaves cleft into rather narrow divisions. Sepals several (10 15), white or purple. Prairie, infrequent.

A Canadensis, L. Stem taller. Leaves larger, cleft into wedge-shaped divisions. Sepals 5, white. Low places, infrequent.

MYOSURUS, L.

Sepals and petals 5. Pistils on a conical receptacle which greatly elongates in fruit.

M. minimus, L. Annual, with narrow, entire, radical leaves. Dry places, frequent.

RANUNCULUS, L.

Sepals and petals 5. Pistils in a head, receptacle not elongating in fruit.

R. abortivus, L. Glabrous, branching, 2-5 dm. high. Lower leaves roundish, crenate; upper parted. Petals small, yellow. Low woods, common.

AQUILEGIA, L.

Sepals 5, colored. Petals 5, strongly spurred below. Pistils 5, many ovuled.

A. Canadensis,, L. A perennial with ternately compound leaves and showy, scarlet and yellow flowers. Rocky woods, not common.

DELPHINIUM, L.

Sepals 5, irregular. one of them spurred. Petals 4, two extending back into the spur of the calyx. Flowers in racemes. Leaves palmately parted. Ours with 3 pistils.

D. Carolinianum, Walt. Racemes wand-like. Flowers light blue or whitish. Lobes of leaves narrowly linear. Prairie, frequent.

D. tricorne, Michx. Raceme rather short. Stem low. Flowers deeper blue. Lobes of leaves broader and wedge-shaped. Rocky woods, infrequent.

24. MENISPERMACEAE.

Woody vines with hypogynous, diœcious flowers.

MENISPERMUM, L.

M. Canadense, L. Leaves roundish, angled, peltate near the edge. Fruit a black drupe with a flat stone. Low woods, common.

25. PAPAVERACEAE.

Sepals 2, small. Petals 4 (or more). Stamens 6 (or more). Pistil one, 1-celled. Herbs

BICUCULLA, Adans.

Corolla 2-spurred. Stamens united in 2 sets of 3 each. Leaves dissected, all radical.

B. Cucullaria, (L.) Millsp. Pinkish flowers in racemes from a granulated bulb. Woods, rare.

CAPNOIDES, Adans.

Corolla 1-spurred. Fruit a many-seeded capsule. Stems leafy. Otherwise like *Bicuculla*.

C. aureum, var. *occidentale*, (Engelm.) Flowers small, yellow, racemed. Woods, common.

26. CRUCIFERAE.

Sepals 4. Petals 4 (or none). Stamens 6, two being shorter, (or only 2). Pistil one, 2-celled by a false partition between the two parietal placentæ. Fruit a 2-valved capsule. Herbs with alternate leaves and white or yellow flowers.

1. Capsule compressed parallel to partition.
   Linear, elongated.
      Leaves dissected, grouped near middle of stem........*Dentaria*
      Leaves undivided; stem leafy............ ....*Arabis*
   Oblong; low plants.......................................*Draba*
2. Capsule terete or 4-angled.
   Short, valves convex, nerveless...........................*Roripa*
   Valves convex or keeled, nerved; flowers very small, yellowish, *Sisymbrium*
   Flowers larger; capsule beaked beyond the valves.........*Brassica*
3. Capsule short, flattened contrary to partition.
   Obcordate..... ....*Bursa*
   Orbicular; stamens only 2; often apetalous...............*Lepidium*

RORIPA, Scop.

R. Armoracia, (L.) The common horse-radish with white flowers and large leaves. Sometimes spontaneous near gardens.

R. Nasturtium, (L.) Rusby. Flowers white. Capsules linear. Leaves pinnate. Wet places, rare.

R. sessiliflora, (Nutt.) Flowers yellowish, very small. Capsules about 1 cm. long, nearly sessile. Leaves lyrate. Wet places, common.

R. sinuata, (Nutt.) Flowers yellow, ½ cm. long. Capsules about 1 cm. long, on slender pedicels. Common.

ARABIS, L.

A. Canadensis, L. Stem ½-1 m. high, leafy. Leaves oblong-lanceolate, not clasping, somewhat toothed. Flowers whitish. Capsules curved, elongated, very flat, pendulous. Woods, frequent.

A. hirsuta, (L.) Scop. Leaves clasping by a sagittate base. Flowers greenish. Capsules upright, straight. Rocky woods, common.

DENTARIA, L.

D. laciniata, Muhl. Flowers white or pink; leaves 3-parted. Woods, rare.

DRABA, L.

D. Caroliniana, Walt. Plants a few cm. high. Flowers white. Leaves obovate, entire, hirsute. Capsules smooth, or in var. *micrantha*, (Nutt.) Gray, minutely hairy. Waste places, common.

SISYMBRIUM, L.

S officinale, (L.) Scop. Leaves 1-pinnatifid. Capsules nearly sessile, closely appressed to stem. A common weed.

S. pinnatum, (Walt.) Greene. Leaves 2-pinnatifid. Capsules spreading on horizontal pedicels. Open ground, common.

BRASSICA, L.

B. nigra, (L.) Koch. Flowers bright yellow, rather large. Leaves with a large terminal lobe and a few small lateral ones. Stem somewhat hairy. Capsules 4-angled. Waste places, frequent.

B. Sinapistrum, Boiss. As in *B nigra*, but stem smooth and glaucous and leaves more or less sagittate-clasping. Capsules terete and knotty. Waste places, frequent.

BURSA, Weber.

B. pastoris, (L.) Wigg. Root leaves clustered, pinnatifid. Stem leaves clasping by a sagittate base. Flowers white, small. A common weed.

LEPIDIUM, L.

L. intermedium, Gray. Low, bushy branched. Leaves narrow, entire. Cotyledons incumbent. Dry places, very common.

L Virginicum, L. Less branched. Leaves toothed. Cotyledons accumbent. Woods, frequent.

L. Draba, L. Pods heart shaped. Leaves sagittate. Escaped from gardens occasionally.

27. CAPPARIDACEAE.

Sepals 4. Petals 4. Stamens 6 or more, equal. Pistil one, 1-celled, with 2 parietal placentæ

CLEOME, L.

Stamens 6. Capsule linear, many seeded, stipitate

C. serrulata, Pursh. Stem smooth, 1 m. high. Leaves 3 foliate. Flowers pink. Waste places, not common.

28. SAXIFRAGACEAE.

TRIBE RIBESIOIDEÆ.

Shrubs with alternate, simple leaves. Fruit a berry.

RIBES, L.

Calyx 5-lobed. Petals 5. Pistil one, adherent to calyx-tube; ovary 1-celled; ovules numerous.

R. gracile, Michx. Stems provided with scattered prickles and stronger triple spines below the leaves. Flowers narrow, whitish. Stamens exerted. Woods, frequent.

29. PLATANACEAE.

Flowers monœcious, in heads, naked. Trees with alternate, palmately lobed leaves and sheathing stipules.

PLATANUS, L.

P. occidentalis, L. Leaves sinuate-toothed. A large tree with white bark peeling off in patches. Near streams, frequent.

30. ROSACEAE.

Sepals 5, united into a cup (often with 5 alternating bractlets). Petals 5, and stamens numerous, inserted on the calyx tube. Leaves alternate, with stipules.

1. Trees or shrubs.
    - Leaves simple.
        - Ovary one, 1-celled, free; fruit a drupe.
            - Flowers in umbels.............................*Prunus*
            - Flowers in racemes.............................*Cerasus*
        - Ovaries 3-5, enclosed in the calyx tube; fruit a pome.
            - Flowers pink, fragrant ....*Pyrus*
            - Flowers white, ill-scented.......... .... *Crataegus*
    - Leaves compound.

Palmate; flowers white..*Rubus*
Pinnate; flowers pink..*Rosa*

2. Herbs.

Flowers white..*Fragaria*
Flowers yellow..*Potentilla*

PRUNUS, L.

P. Americana, Marsh. Flowers appearing before the leaves. Fruit ovoid; stone flat. Leaves ovate, doubly serrate, veiny. Many of the branches thorn-like at apex. A small tree. Thickets, common.

P. angustifolia, Marsh. Resembles the foregoing, but leaves narrower and tending to be partially folded together, finely and evenly glandular-serrate. Fruit globular; stone more turgid. Sand hills along the river. A shrub.

CERASUS, L.

C. Virginiana, (L.) Loisel. Flowers appearing just after the leaves. Fruit and stone globular. Leaves obovate, thin. A tall shrub, not thorny. Woods, not common.

RUBUS, L.

Pistils numerous, becoming small drupelets in fruit. Somewhat woody plants with prickly bark.

R. occidentalis, L. Stems upright or declining, red, glaucous. Leaflets 3, white-downy beneath. Petals shorter than sepals. Thickets and fence corners, frequent.

R. villosus, Ait. Stems upright or declining, brown, furrowed. Leaflets 3-5, the terminal 1-3 stalked, glandular pubescent. Petals longer than sepals. Roadsides, rather uncommon.

R. Canadensis, L. Stems trailing. Leaflets usually 3, thin and nearly smooth. Otherwise resembles the preceding. Woods about St. George.

FRAGARIA, L.

Calyx with 5 bractlets alternating with the lobes. Pistils numerous becoming achenes. The receptacle becoming in fruit enlarged and pulpy. Leaves radical, trifoliate. Flowers cymose on scapes. Spreading by runners.

F. Virginiana, Mill. Low prairie, common.

POTENTILLA, L.

As in *Fragaria* but receptacle not becoming pulpy in fruit. Ours with leafy stem.

P. Monspeliensis, L. Stem erect, 2-5 dm. Leaflets 3. Sandy soil, infrequent.

P. pentandra, Engelm. More slender, flowers smaller and the lower leaves with the lateral leaflets parted nearly to the base, thus appearing 5-foliate. Sandy soil, rare.

ROSA, L.

Calyx tube ovoid, contracted at summit where are borne the petals and stamens, becoming fleshy in fruit. The several pistils at the bottom of the tube becoming bony, hairy achenes.

R. Arkansana, Porter. Stem prickly or nearly unarmed. Prairie, common.

PYRUS, L.

P. Coronaria, L. Twigs and simple leaves glabrous or nearly so. Flowers umbellate. Bluffs along Kansas river, not common. Often thorny with stunted branchlets.

P. Iowensis, (Wood) Bailey. Thicker leaves and the twigs woolly. Woods about St. George.

CRATÆGUS, L.

Twigs bearing smooth, simple thorns.

C. coccinea, L. Leaves ovate-cordate, thin, glabrous, incised. Fruit globose, scarlet. Bluffs, rare.

C. mollis, Scheele. Leaves thicker and larger, downy beneath. Bluffs, rare.

31. LEGUMINOSAE.

Sepals 5, usually more or less united. Petals 5 (one in *Amorpha*), usually irregular. Stamens usually 10, distinct or united. Pistil one, simple. Fruit a legume Leaves alternate, with stipules, mostly compound.

SUBORDER PAPILIONATÆ.

Corolla papilionaceous, the upper petal, or standard, enclosing the others in the bud.

Leaves palmately compound.
  Stamens distinct.. *Baptisia*
  Stamens diadelphous.
    Flowers in heads.. *Trifolium*
    Flowers in spikes or racemes .... *Psoralea*
Leaves pinnately compound (if trifoliate, the terminal leaflet stalked.)
  Leaflets 3.
    Flowers yellow or white..... .... *Melilotus*
    Flowers purple............ .... *Medicago*
  Leaflets more than 3.
    Shrub.. *Amorpha*
    Herbs.
      Leaves odd-pinnate.............................. *Astragalus*
      Leaf stalk terminating in a tendril.................... *Vicia*
      Leaf-stalk terminating in a bristle.......... .... *Lathyrus*

BAPTISIA, Vent.

Stamens distinct. Legume oblong, inflated, stalked in the persistent calyx. Leaves palmately 3-foliate.

B australis, (L.) R. Br. Smooth. Flowers blue in an erect raceme. Prairie, frequent.

B. leucophæa, Nutt. Hairy. Flowers yellow, in a reclining raceme. Prairie, common. (Hybrids with the foregoing occur.)

MEDICAGO, L.

Stamens diadelphous. Leaves pinnately 3-foliate. Legume more or less coiled.

M. sativa, L. Escaped from cultivation, frequent.

MELILOTUS, Juss.

As in *Medicago* but legume not coiled. Flowers in spikes. Tall herbs.

M. alba, Lam. Flowers white. Roadsides, etc., frequent.

M. officinalis. (L.) Lam. Flowers yellow. Waste places, rare.

TRIFOLIUM, L.

Leaves palmately 3-foliate. Flowers in head-like clusters. Low herbs.

T. medium, L. Flowers purple, sessile in the head Leaflets without a dark spot. Fields, common. (*T. pratense*, L., is distinguished by having a dark spot on the leaves.)

T. repens, L. Flowers white (turning pink with age), pedicelled in the head. Leaflets notched at apex. Meadows, common.

PSORALEA, L.

Perennial herbs with glandular dotted, 3-5-foliate leaves and purplish flowers.

P. esculenta, Pursh. Stem 1-3 dm high, rough, hairy. Leaves 5-foliate. Raceme dense. Prairie, not common.

P. tenuiflora, Pursh. Stem taller and much branched, nearly smooth. Raceme loose. Leaves 3-5 foliate. Prairie, common

AMORPHA, L.

Stamens monadelphous only at base. Leaves odd-pinnate. Flowers violet, in spikes. Corolla of one petal.

A. fruticosa, L. Nearly smooth. Leaflets 8-12 pairs Wet places, frequent.

ASTRAGALUS, L.

Stamens diadelphous. Low perennial herbs with racemed flowers. Legume sometimes 2-celled by a false partition.

A. caryocarpus, Ker. Stems leafly and spreading. Leaflets narrowly elliptical (1-1½ cm. long). Flowers about 2 cm. long, purplish, in a short loose raceme. Legume turgid, 2-celled, thick and fleshy, becoming papery at maturity. Prairie, common.

A. lotiflorus, Hook. Stems short and tufted. Leaflets oblong, broader and more distant than in the foregoing. Flowers about 1 cm. long, yellowish, in a head-like raceme. Legume dry, not completely 2-celled. Prairie, frequent.

A. Missouriensis, Nutt. Resembling the foregoing but leaflets smaller and flowers blue or purple. Rocky bluffs, frequent.

VICIA, L.

V. sparsifolia, Nutt. Leaflets 2-5 pairs, linear or oblong. Raceme 2-4 flowered, flowers narrow, about 2 cm. long, narrow. Low ground, not common.

LATHYRUS, L.

L. ornatus, Nutt. Flowers about 3 cm. long, the standard broad and conspicuous, purple. Prairie, frequent.

SUBORDER CÆSALPINIOIDEÆ.

Flowers not papilionaceous, or if somewhat so, the upper petal enclosed by the others in the bud. Stamens usually distinct.

GYMNOCLADUS, Lam.

Flowers regular, whitish, in terminal racemes. Leaves 2-pinnate. Legume oblong, woody.

G. dioica, (L.) Koch. A large tree with nearly diœcious flowers. Low woods, frequent.

GLEDITSCHIA, L.

Flowers small, regular, greenish, in spikes. Legume woody. Thorny trees.

G. triacanthos, L. Leaves 1-2-pinnate. Thorns often much branched, above the axils. Legume elongated. Low woods, frequent.

CERCIS, L.

Corolla imperfectly papilionaceous, pink. Legumes thin and flat. Trees with simple cordate leaves. Flowers appearing before the leaves.

C. Canadensis, L Woods, frequent.

SUBORDER MIMOSOIDEÆ.

Flowers regular, corolla valvate. Stamens distinct, exerted. Leaves 2-pinnate.

SCHRANKIA, Willd.

Corolla gamopetalous, funnel form. Stamens about 10. Legume narrow and prickly.

S. uncinata, Willd A perennial herb with prickly stems, sensitive leaves and axillary, peduncled heads of pink flowers. Prairies, frequent.

32. GERANIACEAE.

Sepals and petals 5. Stamens usually 10. Ovary 5-celled. Herbs with simple but usually deeply lobed leaves.

GERANIUM, L.

G. Carolinianum, L. A low much branched herb with 5-parted and much cleft leaves, and small pinkish flowers. Sterile soil, frequent.

33. OXALIDACEAE.

Differs from the preceeding order in having compound leaves, in ours 3 foliate, with obcordate leaflets.

OXALIS, L.

O. corniculata, L. Stem leafy, from a creeping rootstock. Flowers yellow. Fruit erect on a reflexed pedicel. Open ground, common.

O. stricta, L Differs from the above in having the fruiting pedicels ascending. Common.

O. violacea, L. Leaves radical, from a scaly bulb. Flowers pink or purple. Rocky soil, common

34. RUTACEAE.

ZANTHOXYLUM, L.

Flowers diœcious. Petals and stamens 4-5, the latter alternate with petals. Pistils 2 6. Fruit a fleshy. 2-valved, 1-2-seeded capsule. Prickly shrubs with pinnate, glandular dotted leaves and small greenish flowers.

Z Americanum. Mill. Calyx none. Plant aromatic. Prickles in position of stipules and sometimes also scattered. Leaflets 3-5. Woods, common.

35. EUPHORBIACEAE.

Flowers monœcious or diœcious, mostly apetalous. Ovary free, 3-celled, 3-ovuled. Fruit a capsule.

EUPHORBIA, L.

Flowers monœcious, naked, enclosed in a cup-shaped involcure, the whole likely to be mistaken for a single flower. Staminate flowers, several in each involucre, consisting of a single stamen jointed on its pedicel. Pistillate flower one, pedicelled. Involucre often provided with 4-5 thick glands with colored appendages, making the resemblance to a flower still more striking. Herbs with milky juice.

E. Cyparissias, L. Erect perennial with linear, entire leaves scattered along the stem, the upper rounded. Inflorescence umbellate. Escaped from gardens.

E. nutans, Lag. Annual. Stem ascending or erect, slender, nearly smooth. Leaves oblong, serrate, opposite. Open places, common.

E. maculata, L. Prostrate, hairy annual. Leaves oblong, often with a brown spot. Open places, common.

TRAGIA, L.

Flowers apetalous, monœcious, racemose. Calyx present.

T. stylaris, Muell. An erect perennial, hispid, with stinging hairs. Leaves ovate to oblong, coarsely toothed. Dry hills, rare.

36. ANACARDIACEAE.

Sepals, petals and stamens 5. Ovary 1-celled, 1-ovuled. Styles and stigmas 3.

RHUS, L.

Shrubs with compound leaves and small greenish or yellowish flowers.

R. aromatica, Ait. Flowers in compact clusters which develope before the leaves. A low, upright shrub. Leaves 3-foliate. Rocky bluffs, frequent.

R. radicans, L. Flowers in loose axillary panicles, developing after the leaves. An upright shrub or high climbing vine. Leaves 3-foliate. Woods and thickets, common.

37. CELASTRACEAE.

Sepals, petals and stamens 4-5, the latter alternate with petals and inserted on a disk surrounding the ovary. Shrubs with simple leaves.

CELASTRUS, L

Flowers greenish yellow, in terminal racemes. Leaves alternate. Fruit globose.

C. scandens, L. A twining vine, with ovate or oval, finely serrate leaves. Upland woods, frequent.

EUONYMUS, L.

Flowers dark red in axillary clusters. Leaves opposite. Fruit lobed.

E. atropurpureus, Jacq. An upright shrub with 4-sided branches. Parts of flower usually in fours. Woods, frequent.

38. STAPHYLEACEAE.

Flowers perfect, regular. Stamens as many as petals. Shrubs with opposite, compound leaves. Ovary 3-celled.

STAPHYLEA, L.

Petals 5 Ovules several. Fruit much inflated.

S. trifolia, L. Leaves pinnately 3-foliate. White flowers in drooping racemes. Woods, not common.

39. ACERACEAE.

Flowers polygamous or diœcious, regular. Ours apetalous. Ovary 2-celled and 2-lobed, 4-ovuled. Fruit a samara (double), unsymmetrically winged at apex. Leaves opposite. Woody plants.

ACER, L.

Leaves simple, palmately lobed.

A. saccharinum, L. Flowers in umbel-like clusters much preceding the leaves. Common along Blue river.

RULAC, Adans.

Leaves pinnately 3-5-foliate. Flowers diœcious, the staminate on capillary drooping pedicels.

R. Negundo, (L) Hitchc. Low woods, frequent.

40. HIPPOCASTANACEAE.

Flowers usually irregular. Stamens more numerous than petals. Ovary 3-celled. Woody plants with compound leaves.

ÆSCULUS, L.

Calyx tubular. Petals unequal. Stamens 7. Leaves palmate.

Æ. arguta, Buckl. A small tree with polygamous, yellow flowers in terminal panicles appearing with the leaves. Petals 4. Leaflets 7. Woods, frequent.

41. RHAMNACEAE.

Sepals, petals and stamens 4-5, the latter opposite the petals. Shrubs with alternate, simple leaves.

RHAMNUS, L.

Disk lining the calyx free from ovary. Fruit a drupe.

R. lanceolata, Pursh. Flowers more or less clustered, axillary, greenish, the parts in four's. Upland woods, rare.

CEANOTHUS, L.

Disk lining the calyx cohering with the ovary. Fruit dry, separating into 3 parts at maturity.

C. ovatus, Desf. A low shrub with gladular-serrate leaves and white flowers in dense, terminal, umbel-like clusters. Prairie and open woods, common.

42. VITACEAE.

Sepals (when present), petals and stamens 4-5, the latter opposite the petals Ovary 2-celled, 4-ovuled Woody vines provided with tendrils opposite the leaves. Fruit a berry.

VITIS, L.

Calyx lobes scarcely developed. Petals separating below and falling from the expanding flower without opening. Leaves simple. Flowers in a compound panicle, small and very fragrant.

V. vulpina, L. Leaves cordate, 3-lobed, incised-serrate, smooth. Woods, common.

PARTHENOCISSUS, Planch.

Calyx shortly 5-toothed. Petals expanding. Leaves palmately compound.

P. quinquefolia, (L.) Planch Leaflets 5, clinging to rough supports by disk-like terminations. Woods, common. The older parts provided with aerial rootlets.

P. vitacea, (Knerr) Hitchc. Differs from the above in having no aerial rootlets; the canes smooth, and lighter colored; the tendrils dichotomous like the grape, and usually without disks; the inflorescence dichotomous rather than pinnate as in the former; flowering about two weeks earlier; the fruit maturing earlier and considerably larger. Woods, infrequent.

43. MALVACEAE.

Sepals 5, more or less united, often provided with a whorl of bractlets outside. Petals 5. Stamens numerous, united in a column. Pistils several, united in a ring. Leaves alternate, with stipules. Ours herbs.

MALVA, L.

Involucel of 3 bractlets. Carpels rounded, beakless.

M. rotundifolia, L. Stems spreading. Leaves on long petioles, round-cordate, crenate. Petals whitish. Waste places, infrequent.

CALLIRHOE, Nutt.

Involucel 3-leaved or none. Carpels beaked.

C. alcæoides, (Michx.) Gray. Stem erect. Involucel none. Flowers pink or white. Prairie, not common.

C. involucrata, (Nutt.) Gray. Stem spreading. Involucel 3-leaved. Flowers deep red. Prairies, common.

44. VIOLACEAE.

Sepals 5. Petals 5, irregular. Stamens 5. Pistil one, 1-celled, with 3 parietal placentæ. Herbs.

VIOLA, L.

Sepals eared at base. The lower petal spurred and two of the stamens sending appendages into the spur. Ours with blue flowers.

V. obliqua, Hill. Stemless. Leaves cordate, crenate. Woods, common.

V. pedatifida, Don. Stemless. Leaves palmately cleft into narrow lobes. Prairie, frequent.

V. tenella, Muhl. Stem low, with oval leaves and large pinnatifid stipules. Flowers small, light blue. Grassy places, frequent.

CALCEOLARIA, Lœfl.

Sepals not eared. Leafy perennials with small axillary flowers.

C. verticillata, (Ort.) Kuntze. Leaves linear. Flowers white. Prairie, infrequent.

45. OENOTHERACEAE.

Parts of flower usually in four's. Pistil one, adherent to calyx tube, 4-celled. Herbs.

ŒNOTHERA, L.

Calyx tube prolonged beyond the ovary, lobes 4. Petals 4. Stamens 8.

Œ. Missouriensis, Sims. Calyx tube about 1 dm. long. Flowers axillary, ½-1 dm. broad, yellow. Low decumbent perennials with silky pubescence and broadly 4-winged capsules. Rocky hills, not uncommon.

Œ. serrulata, Nutt. Stems slender, from a woody base. Leaves linear, denticulate. Flowers axillary, yellow, 1½-2 cm. wide. Capsule narrow, cylindrical, not winged. Prairie, frequent.

Œ. sinuata, L. Stems hairy, decumbent. Leaves lanceolate, sinuately toothed or pinnatifid. Flowers yellow. Capsule cylindrical, hairy. Prairie, frequent.

Œ. speciosa, Nutt. Stem erect. Leaves more or less pinnatifid. Flowers white or pinkish. Capsule spindle shaped, strongly ribbed. Prairie, frequent.

GAURA, L.

Flowers much as in *Œnothera*, but in our species small and in loose spikes. Fruit indehiscent and nut-like, 4-angled.

G. coccinea, Pursh. A canescent leafy perennial with linear leaves and pink or scarlet flowers. Prairie, common.

46. UMBELLIFERAE.

Calyx adhering to the ovary, lobes minute or wanting. Petals and stamens 5, inserted on a disk that crowns the 2-celled, 2-ovuled ovary. Styles 2. Fruit separating into 2 seed-like carpels. Herbs with alternate, usually compound leaves, the base of the petiole expanding and clasping around the stem. Flowers small in simple or compound umbels.

Stemless; perennial from a thick root........................*Peucedanum*
Leafy stemmed.
  Annual; leaves finely dissected..*Apium*
  Perennial; leaflets ovate or oblong.
    Flowers capitate in umbellet..........................*Sanicula*
    Flowers long pedicelled in umbellet.
      White..*Osmorrhiza*
      Yellow..*Polytænia*

SANICULA, L.

Fruit globose, not flattened, densely prickly with hooked bristles. Stems about 1 m. high, glabrous. Leaves palmately parted. Flowers perfect with staminate ones intermixed.

S. Canadensis, L. Staminate flowers few and short pedicelled. Flowers whitish. Low woods, common.

S. Marilandica, L. Staminate flowers more numerous and long pedicelled. Flowers yellow. Low woods, common.

APIUM, L.

Fruit ovate or roundish, not flattened. Flowers white.

A. patens, (Nutt.) Wats. A slender, divaricately branched annual (3-6 dm.), with leaves finely dissected into filiform lobes, and tuberculate fruit. Prairie, frequent.

OSMORRHIZA, Raf.

Fruit narrow, not flattened, attenuate at base, bristly. Flowers white.

O. longistylis, (Torr.) DC. Leaves ternately compound, leaflets ovate, incised. Stem ½-1 m. Woods, not common.

PEUCEDANUM, L.

Fruit ovate, flattened parallel to partition, ribbed on back, winged on margin.

P. fœniculaceum, Nutt. Leaves much dissected. Flowers yellow. Plant with a strong odor resembling celery. Rocky bluffs, frequent.

POLYTÆNIA, DC.

Fruit oval, flattened parallel to partition, with corky margins. Flowers yellow. Leaves 2-pinnate, leaflets incised.

P. Nuttallii, DC. ½-1 m. high. Rocky soil, frequent.

47. CORNACEAE.

Calyx adherent to ovary. Petals and stamens inserted on disk which crowns the ovary. Style one. Fruit a drupe. Woody plants.

CORNUS, L.

Petals and stamens 4. Ovary 2-celled, 2-ovuled. Leaves opposite. Flowers (in ours) white in flat topped cymes.

C. asperifolia, Michx. Branches brown or gray. Leaves rough, pubescent above, downy beneath. Fruit white. An upright shrub or small tree. Thickets, common.

C. sericea, L. Branchlets bright red-brown, woolly pubescent. Leaves elliptical, whitened and nearly glabrous beneath. Branches often elongated, declining and stoloniferous at summit. Fruit drab or bluish. Wet places, not common.

48. PRIMULACEAE.

Calyx free. Corolla 4-5-lobed. Stamens as many as corolla lobes and inserted opposite them. Ovary 1-celled, many ovuled, with central placenta. Herbs with simple leaves.

ANDROSACE, L.

Low plants with radical leaves and mostly umbelled flowers. Capsule 5-valved.

A. occidentalis, Pursh. Annual. Scapes several. Corolla white, small. Sterile hills, common.

49. OLEACEAE.

Calyx and corolla 4-cleft, one or both sometimes absent. Stamens 2. Ovary 2-celled. Woody plants, with opposite leaves.

FRAXINUS, L.

Flowers diœcious. Calyx very small. Corolla absent. Fruit a samara, winged at apex. Trees with pinnate leaves. Flowers in panicles on the old wood.

F. viridis, Michx. f. A small tree with glabrous twigs, or in var. *pubescens*, Hitchc., the twigs pubescent. Woods, common.

50. APOCYNACEAE.

Calyx and corolla 5-lobed. Stamens 5, inserted on corolla tube, alternate. Ovaries 2, free from calyx. Plants with milky juice and entire, opposite leaves.

APOCYNUM, L.

Fruit 2 long and slender follicles. Flowers white, cymose.

A. cannabinum, L. About 1 m. high. Leaves oval. Flowers small, in close, terminal cymes. Low ground, frequent.

51. ASCLEPIADACEAE.

Characters about as in the last order but anthers connected with stigma. Leaves opposite or alternate.

ASCLEPIODORA, Gray.

Corolla lobes 5, ascending. Inside there is a crown consisting of 5 spreading, hooded bodies. Flowers in umbels.

A. viridis, (Walt) Gray Stem low. Leaves alternate. Flowers 2-3 cm. broad, green with purple crown. Prairie, not common.

ACERATES, Ell.

Corolla lobes 5, reflexed. Crown present. Flowers small, greenish, in compact umbels.

A. lanuginosa, (Nutt.) Dec. A low, hairy perennial with opposite or scattered leaves and a single umbel terminating the stem. Prairies, rare.

52. CONVOLVULACEAE.

Sepals 5. Corolla 5-lobed, covolute. Stames 5, inserted on the corolla tube alternating with its lobes. Ovary 2-celled, 4-ovuled. Herbs with alternate leaves and erect or usually twining stem. Fruit a capsule.

CONVOLVULUS, L.

Corolla funnel form, showy. Style one.

C. Sepium, L Stem usually twining, glabrous. Leaves halberd-shaped. Corolla white, 4-5 cm long. Moist soil, common.

C repens, L Stem usually trailing, pubescent. Corolla pinkish or white. Low ground, frequent.

EVOLVULUS, L.

Corolla small, nearly rotate. Styles 2.

E pilosus, Nutt. A low, erect, silky hairy herb with crowded, narrow leaves and small, blue flowers (½ cm. broad). Prairie, frequent.

53. POLEMONIACEAE.

As in the preceding order but ovary 3-celled. Herbs, not twining.

PHLOX, L.

Corolla salver-form. Perennials with opposite, entire, sessile leaves and cymose inflorescence.

P. divaricata, L Stems ascending from a decumbent base. Leaves oblong to ovate. Corolla pale blue. Woods, frequent.

P. pilosa, L. Stem erect. Leaves linear. Corolla pink. Prairie, frequent.

54. HYDROPHYLLACEAE.

As in the foregoing orders but ovary 1-celled with 2 parietal placentæ. Styles 2, more or less united. Herbs.

MACROCALYX, Trew.

Corolla small, bell shaped. Placentae meeting in the axis so as to make the ovary appear 2-celled.

M. Nyctelea, (L.) Kuntze. A low branched annual with pinnately parted leaves and nearly white flowers solitary in the forks or opposite the leaves. Low woods, common.

55. ASPERIFOLIAE.

As in the foregoing orders but ovary 4-lobed around the base of the single style, separating at maturity into 4 seed-like nutlets Hairy herbs with alternate, entire leaves and flowers in one sided racemes coiled at the tip.

CYNOGLOSSUM, L.

Throat of corolla closed by 5 obtuse scales. Nutlets armed all over with short barbed prickles

C. officinale, L Corolla dull red Lower leaves petioled, upper sessile. Waste places, frequent.

LAPPULA, Moench.

Much as in *Cynoglossum* but nutlets armed only on the back or edges.

L. pilosa, (Nutt.) An erect, diffusely branched, hispid annual, with narrow leaves and small, pale blue flowers Nutlets armed with a single row of prickles on margin. Dry hills, common.

LITHOSPERMUM, L.

Throat of corolla more or less closed. Nutlets whitish, bony, smooth or roughish. Ours with deep perennial roots.

L. angustifolium, Michx A few dm. high, rough hairy. Leaves linear. Flowers lemon yellow, with tube 2-4 times as long as calyx Prairie, frequent

L. canescens, (Michx.) Lehm. Soft hairy, low. Leaves oblong. Corolla orange, tube only a little longer than calyx. Prairies, frequent.

L Gmelini, (Michx.) Stem taller ($\frac{1}{3}$-$\frac{1}{2}$ m.), rough hairy. Leaves oblong to linear. Corolla orange, tube short. Dry hills, frequent.

ONOSMODIUM, Michx.

Corolla tubular, throat open, lobes triangular, scarcely opening. Nutlets bony.

O. molle, Michx. About 1 m. high, coarsely hairy. Leaves elliptical, strongly ribbed. Flowers whitish, with much exerted style. Prairie, common.

56. VERBENACEAE.

Flowers irregular. Stamens 4. Ovary 2-4-celled, splitting at maturity into as many nutlets. Leaves opposite. Ours all herbs.

VERBENA, L.

Calyx tubular. Corolla 5-lobed, salver-form. Flowers in spikes.

V. Aubletia, L. Stem low. Leaves ovate, incisely lobed. Corolla pinkish, 1-1½ cm. broad. Spikes depressed, elongating in fruit Bracts narrow. Prairie, frequent

V. bracteosa, Michx. Procumbent. Bracts large, pinnatifid Flowers blue, small, shorter than the bracts. Prairie, common.

57. LABIATAE.

Corolla 2-lipped (sometimes nearly regular). Stamens 4 or 2. Ovary 4-lobed and separating into nutlets as in *Asperifoliæ*. Herbs with square stems, opposite leaves and an aromatic odor.

Stamens 2.

Calyx tubular, lobes awl-shaped.... ....*Hedeoma*

Calyx strongly 2-lipped, lobes broad ....*Salvia*

Stamens 4.

Calyx lobes spiny toothed.

Teeth 10 ....*Marrubium*

Teeth 5................ ....*Leonurus*

Calyx lobes not spiny.

Tubular, 5-toothed........ ... .*Glechoma*

Strongly 2-lipped.... ....*Scutellaria*

HEDEOMA, Pers

Calyx gibbous, nerved. Low annuals, with small leaves.

H hispida, Pursh. Leaves narrow, entire. Corolla very small, blue. Prairie, common.

SALVIA, L.

S lanceolata, Willd 2 3 dm. high. Leaves narrow, entire or nearly so. Inflorescence apparently interrupted-spicate. Corolla blue, 1 cm. long. Prairie, common.

GLECHOMA, L.

G. hederacea, L Creeping. Leaves rounded, crenate. Corolla blue. Yards and roadsides, rare.

SCUTELLARIA, L.

Upper lip of calyx helmet-shaped.

S parvula, Michx. Low, 1-2 dm. Leaves ovate. Flowers in the axils of the upper leaves. Corolla blue, about 1 cm. long Prairie, common.

MARRUBIUM, L.

M vulgare, L Leaves rounded, crenate, petioled. Flowers in heads. Corolla white, small. Waste places, infrequent.

LEONURUS, L.

L. Cardiaca, L Tall. Leaves cut lobed Flowers pink. Waste places, infrequent.

58. SOLANACEAE.

Corolla 5-lobed. Stamens 5, inserted on corolla tube, alternate. Ovary 2 celled, many ovuled. Herbs with alternate leaves.

SOLANUM, L.

Corolla rotate. Stamens exerted Fruit a berry. Ours prickly stemmed

S. Carolinense, L. Stellate hairy. Leaves, sinuate lobed Flowers pale bule or nearly wl ite. Perennial, sandy ground, frequent.

S rostratum, Dunal. Leaves deeply 1-2-pinnatifid. Flowers yellow. Berry enclosed in the prickly calyx. Annual. Common in summer and autumn.

PHYSALIS, L.

Corolla funnel form with a dark "eye," fruit a berry, enclosed in the enlarged and papery calyx. Ours perennials with yellow flowers.

P. lanceolata, Michx. 3-4 dm. high, hirsute. Leaves lanceolate, nearly entire. Dry ground, common.

P. longifolia, Nutt. Taller, glabrous. Leaves larger and often sinuate toothed. Prairie, frequent.

P. cinerascens, (Dunal). Low, villous. Leaves usually entire. Sandy soil, frequent.

P. Virginiana, Mill. Diffusely branched; viscid with glandular hairs. Leaves ovate, cordate, sinuate toothed. Sandy soil, common.

59. SCROPHULARIACEAE.

Flowers 2-lipped or oblique (nearly regular in *Veronica*). Stamens 4 or 2. Ovary 2-celled, many ovuled Fruit a capsule. Herbs, ours with opposite leaves.

Flowers small, nearly regular, in spikes Stamens 2.... .... *Veronica*
Flowers 2-lipped. Stamens 4.
  Yellow.............. ....*Mimulus*
  Not yellow
    Large, white to pink or blue..........................*Pentstemon*
    Small, lurid within, green without.......... ....*Scrophularia*

SCROPHULARIA, L.

Corolla tube globular, lobes short, 4 upper erect, lower spreading. A scale at summit of tube represents a fifth stamen. Flowers in a terminal panicle.

S. Marilandica, L. 1 m. high or more, erect, smooth. Leaves ovate, serrate. Low woods, common.

PENTSTEMON, Soland.

Corolla tubular below, inflated above. Sterile stamen nearly as long as the others. Flowers showy in an elongated panicle. Erect perennials.

P. Cobæa, Nutt. Clammy-hairy. Leaves ovate or oblong, serrate. Corolla white or pinkish, 4-5 cm. long. Rocky hills, not uncommon.

P. grandiflorus, Nutt. Glabrous and glaucous. Upper leaves rounded and clasping, radical obovate, all entire. Flowers bluish, 4-5 cm. long. Sandy soil, not common.

MIMULUS, L.

Calyx 5 angled. Corolla tubular. Flowers axillary.

M. Jamesii, Torr. & Gray. Creeping at base. Leaves rounded, dentate. Flowers 2-3 cm. long. Ditches and springs, not common.

VERONICA, L.

Calyx 4-parted. Corolla rotate or salver-form. Capsule flat, notched at apex.

V. peregrina, L. Low annual with small, narrow leaves and white flowers in an elongated, terminal spike. Low ground, common.

60. PLANTAGINACEAE.

Sepals 4. Corolla 4-lobed, dry and papery. Stamens mostly 4, inserted on corolla tube. Leaves radical. Flowers in spikes.

PLANTAGO. L.

Ovary 2-celled, ovules few. Leaves ribbed.

P. gnaphalioides, Nutt. Leaves narrow, white with silky hairs. Spike narrow, 1-10 cm. Peduncle about 1 dm. Sterile soil, common.

P. lanceolata, L. Leaves lanceolate. Spike globose or oblong, raised on a long (¼ m. or less) peduncle. Meadows, not common.

61. RUBIACEAE.

Calyx adherent to the ovary. Stamens as many as lobes of corolla and inserted on its tube. Leaves opposite, with stipules, or whorled.

GALIUM, L.

Corolla 4-parted. Ovary 2-celled, 2-ovuled, separating at maturity into 2 seed-like carpels. Herbs, with whorled leaves, square stems and small flowers.

G. Aparine, L. A weak, slender annual, with stems retrorsely bristly on the angles. Leaves 8 in a whorl, linear. Flowers greenish. Fruit bristly with hooked prickles.

62. CAPRIFOLIACEAE.

Differs from the preceding order in having the leaves opposite without stipules. Mostly shrubs but ours an herb.

TRIOSTEUM, L.

Calyx lobes 5, linear, persistent, corolla tubular, somewhat irregular, lobes 5. Ovary 3-celled. Fruit 3-seeded, somewhat fleshy. Leaves connate around the stem, the flowers sessile in the axils.

T. perfoliatum, T. Flowers clustered, dark red. Fruit orange. Woods, frequent.

63. CAMPANULACEAE.

Calyx adherent to ovary. Corolla 5-lobed. Stamens 5, free from corolla, and distinct Flowers blue. Herbs with milky juice and alternate leaves.

CAMPANULA, L.

Capsule opening by 3-5 holes in the sides

C. perfoliata, L. Leaves rounded, clasping, toothed. A low, simple-stemmed annual with axillary flowers. Open ground, common.

64. CICHORIACEAE.

Flowers in heads surrounded by one or more rows of bracts forming an involucre. Calyx adherent to ovary, lobes absent or represented by scales, bristles or teeth (the pappus) which crown the ovary. Corrolla strap-shaped. Stamens 5, inserted on tube of corolla Ovary 1-celled, 1-ovuled. Fruit an achene. Herbs with milky juice and alternate leaves.

NOTHOCALAIS, Greene.

Bracts of involucre erect in 2-3 rows. Leaves radical, linear. Scape bearing a single large head of yellow flowers. Pappus of capillary bristles.

N. cuspidata. (Pursh) Greene. Achenes beakless. Prairie, frequent.

TARAXACUM, Hall.

Differs from *Nothocalais* in having two distinct series of involucral bracts, the outer of short scales. Leaves pinnatifid. Achenes beaked

T. Taraxacum, Karst. Outer involucre reflexed. Waste places, not common.

65. COMPOSITAE.

Differs from the previous order in having all or all except the outer row of corollas. tubular and 5-lobed. The outer row of flowers are often provided with strap-shaped corollas and are then called ray flowers, the central portion of the head being called the disk. Ours all herbs. Leaves opposite or alternate.

Ray flowers none.
- Leaves entire; heads diœcious........................... *Antennaria*
- Leaves dissected; flowers perfect. .. .. *Hymenopappus*

Rays present. but sometimes quite small.
- Yellow.
  - Leaves alternate .... *Senecio*
  - Leaves opposite..... .... *Dysodia*
- White or nearly so.
  - Rays only 4-5; heads numerous in a corymb. .... *Achillea*
  - Rays numerous; heads few or single.
    - Perennial; head solitary; leaves pinnatifid... *Chrysanthemum*
    - Annual; heads few; leaves mostly entire .... *Erigeron*

ERIGERON, L.

Bracts narrow, equal, and in about one row. Pappus of capillary bristles.

E. ramosus, (Walt.) B. S. P. Leaves narrow, sessile, entire or the lower toothed. Heads few in a loose corymb. Stem erect, about ½ m. Fields and prairie, common.

ANTENNARIA, Gærtn.

Bracts dry and papery, imbricated. Pappus of capillary bristles. Woolly perennials.

A. plantaginifolia, (L.) Richards. Low and spreading. Radical leaves obovate: those of the scape-like flower stems small and scale-like. Prairie. common in early spring.

HYMENOPAPPUS, L'Her.

Bracts broad and thin, the upper part somewhat colored. Pappus a row of thin scales. Leaves alternate. Heads several in a loose corymb.

H. corymbosus, Torr. and Gray Erect, about ½ m. high, woolly but becoming glabrous. Upper part of stem leafless Rocky bluffs, not common.

DYSODIA. Cav.

Bracts in one row, united into a cup, some loose ones at base. Pappus a row of chaffy scales, dissected into bristles. Conspicuously dotted with yellow glands giving off a strong odor.

D. papposa, (Vent.) Hitchc. A low annual with pinnately parted leaves and small heads with few, short and inconspicuous rays. Sterile ground, common late in the season.

ACHILLEA, L.

Bracts imbricated, margins scarious. Pappus none.

A. millefolium, L. Leaves alternate, 2-pinnately parted into narrow divisions. Perennial. Prairie, common.

CHRYSANTHEMUM, L.

Bracts imbricated. Receptacle broad and flat. Pappus none.

C. Leucanthemum, L. Stem erect, naked above. Fields, not common.

SENECIO, L.

Bracts equal, in one row. Pappus of capillary bristles.

S. Balsamitæ, Muhl. Erect, woolly. Leaves pinnatifid. Heads several, corymbose. Open ground, common.

www.ingramcontent.com/pod-product-compliance
Lightning Source LLC
Chambersburg PA
CBHW022033190326
41519CB00010B/1696

* 9 7 8 3 3 3 7 2 6 8 5 3 4 *